© 1990 Franklin Watts

Franklin Watts Inc.
387 Park Avenue South
New York, NY 10016

Library of Congress Cataloging-in-Publication Data

Watts, Barrie.
  24 hours in a forest/Barrie Watts.
    p.   cm.   (24 hours)
  Summary: Relates what happens to the plants and animals of the forest during a twenty-four hour period.
  ISBN 0-531-14036-9
  1. Forest ecology—Juvenile literature.   [1. Forest ecology.]
I. Title. II. Title: Twenty-four hours in a forest. III. Series.
QH541.5.F6W37   1990
574.5'2642—dc20   89–38986
                                                          CIP AC

Editor: Hazel Poole
Design: K and Co
Consultant: Michael Chinery

Printed in Belgium
All rights reserved

24 HOURS IN A FOREST

Text and photography by Barrie Watts

FRANKLIN WATTS

NEW YORK · LONDON · SYDNEY · TORONTO

CONTENTS
- Early Morning — 7
- Daytime — 15
- Evening — 25
- Night — 33

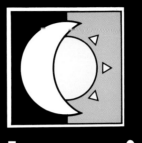

Early Morning

A new day dawns,
and some of the early morning creatures begin to emerge .

Early Morning

The forest slowly awakens to the start of a new day and all the nocturnal animals have now returned to their homes and hiding places. As the sky brightens, the first rays of sunlight penetrate the depths of the dark forest. Some of the animals that have slept throughout the night make their first excursions of the day to search for food or to proclaim their territories.

The birds of the forest are some of the first creatures to wake up. Robins and blackbirds are among the first to start singing. The singing becomes stronger during the breeding season when the birds mark the boundaries of their territories, warning other birds to keep away. As the warming sun shows itself above the horizon, the singing dies down. It is the signal for the start of the daily activity in the forest.

▲ The gray squirrel is the acrobat of the forest. It is able to move through the forest tree canopy at great speed, jumping from branch to branch. It feeds on many things, seeds, fruits, nuts and flowers. It gnaws tree bark and even eats bird's eggs and insects.

◀ The British robin is one of the most common birds in the forest. It is a very aggressive, territorial bird, especially towards its own kind. It is one of the first birds to start the dawn chorus.

As the air cooled during the night, a mist developed and dew drops formed on the leaves and grass. The dew provides a supply of drinking water for beetles and many other small insects and forest animals, especially in the summer when rain is scarce. As the sun rises, the warmth of the air begins to evaporate the mist and the dew.

Squirrels are among the first of the larger mammals to be seen in the early morning. Before they start their constant search for fruits and nuts, they clean and preen themselves, often sunbathing in a golden shaft of warm sunlight.

▲ The red deer stag spends most of the summer apart from the females and young deer. Towards the end of the summer, when the rutting season begins, he will begin to look for a mate. Red deer feed on grass, leaves, bark, mushrooms and lichens.

Early Morning

During the night, many insects were hidden among the foliage for protection. They, too, are covered with the early morning dew and are unable to fly. They now start to crawl out into the sunshine to warm up.

The first to emerge are the flies and butterflies. Within minutes, the dew has gone, flight muscles have been warmed and they are now able to fly.

At the edge of the forest pool, young dragonflies, called nymphs, are crawling up the reeds. They have spent a year or more in the water, but now they are ready for life in the air. Their skins split open and out come the adult dragonflies. After an hour or two in the sun, the insects are ready to fly off and feed.

▶ The early morning dew on the ruddy darter's wings means that it is not able to fly. Within an hour or two, however, after it has warmed up, this small dragonfly will be chasing after its first meal of the day.

▼ The tortoiseshell butterfly warms itself in the early morning sunlight before flying off to a sunnier spot in the forest.

◀ This small fly has spent the night among the flowers of the rosebay willowherb. As the sun rises, it will evaporate the water droplets on its body.

14 Early Morning

▶ In the early morning, the damp mist and dew turns many spiders' webs into rows of glistening jewels. Each kind of spider makes its own particular type of web. This spider has built an orb web (below). Other spiders build hammock and sheet webs.

(Below right) The nymph of the southern hawker dragonfly climbs a reed in the early morning. Within an hour, the adult dragonfly emerges and pumps up its wings. It will take a few days to get its full colors.

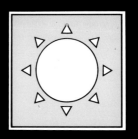

Daytime

As the sun rises above the trees,
life below becomes even busier.

18 Daytime

As the sun rises, small mammals, like the shrew, start to appear. Shrews have a high metabolic rate and are constantly looking for food. They need to feed regularly and sleep for short periods at a time before rushing off to find another meal.

Bank voles are good climbers and can climb high, prickly brambles to search for juicy blackberries to eat. Although these small mammals feed during the day, they are very careful to keep hidden for fear of being eaten themselves. They use the cover of thick undergrowth for safety.

Around midday, baby foxes appear above ground to rest and play. They are very cautious and smell the air and listen for danger before emerging.

▼ This pygmy shrew is eating a woodlouse. It weighs only $1/8$ oz and is the smallest mammal living in the forest. (Below left) The bank vole mainly eats fruits, nuts and seeds.

▼ This young fox is cautiously making sure that it is safe before it leaves the security of its home. It will sunbathe and play with its brothers and sisters in the warm sunlight. Should it be disturbed, it will rush back to its burrow and wait until the evening before coming out again.

Daytime

The temperature of the forest is at its highest in the early afternoon. Amphibians, like frogs and newts, must not dry out so they stay close to the forest pool where the damper conditions make an ideal habitat for them.

Beetles often climb a plant or grass stem to gain height in order to help them get airborne. Because they have to unfold their wings from underneath the wing cases (elytra), they are unable to take flight instantly. The majority of them, therefore, prefer to scuttle over the forest floor in search of their food.

▼ In the heat of the early afternoon sun, this spotted longhorn beetle is busily feeding on the nectar of the bramble flowers growing in the depths of the forest. It is a slow, clumsy flier and will only take flight if the weather is warm.

◄ Common frogs live near to the forest pool which they use to keep cool during the day, and especially in the hot summer months. By then, the pool will be alive with tiny froglets which have developed from the mass of spawn laid in the spring.

Also on the forest floor, the worker carpenter ants are seeking food and carrying materials to and from their dome-shaped nest. The temperature inside the nest is kept at a constant 77°F so that the eggs and larvae develop properly. Each day in mid-summer, the worker carpenter ants take approximately half a pound of sugar to the nest to feed to their larvae. Carpenter ants are good climbers and will climb to the top of the highest tree in the forest in order to collect sugary honeydew from aphids feeding on twigs and branches.

▼ In the early afternoon, male and queen carpenter ants swarm out of their nests and fly through the trees. Mating usually takes place on the ground after this flight. This queen is about to take off. After mating, she will shed her wings and move to another nest.

22 **Daytime**

▲ The brown speckled wood butterfly lives in the open glades of the forest as well as in the darker areas. It flies from leaf to leaf basking in the sunlight. The common blue butterfly (Inset) needs open grassy areas in which to live and breed. It lays its eggs on clover leaves and usually flies close to the ground, sipping nectar from low-growing flowers.

The afternoon is the best time for butterflies to feed on the flowers of the forest. By then, the flowers will have warmed enough to produce good quantities of sweet nectar. When a butterfly settles on a flower, it uncurls its long tongue and probes between the petals to find the nectar. The tongue is hollow and the butterfly uses it to suck up the energy-giving nectar.

Flying from flower to flower, the peacock butterfly feeds until it has taken sufficient nectar. Then it will bask in the late afternoon sun, its fully-open wings facing the last warm rays of sunlight.

▲ The peacock butterfly likes to bask in the warm afternoon sunlight. Its open wings take full advantage of the sun. When they are closed, the dark undersides look like dead leaves, so it is well camouflaged.

▶ During the day, the garden tiger moth rests on old wood and tree bark, usually hidden by the forest vegetation. It is not easily seen and will wait until the evening to fly off and show its bright red underwings.

Daytime

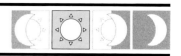

The cool evening is approaching and the female southern hawker dragonfly carefully pushes her long thin eggs into a damp moss-covered log by the waterside. The eggs will stay here until they hatch in the spring. The tiny nymphs that emerge will crawl into the water where they will spend two or three years, feeding on tadpoles and other small water creatures until they are ready to crawl out one morning and change into adult dragonflies.

▶ This female southern hawker dragonfly is pushing her eggs into a log by the forest pool. They will hatch the following spring, and the nymphs will crawl into the water.

◀ Throughout the day, the small ruddy darter dragonfly sunbathes by the waterside. It rests on the same spot every day and only darts off to chase other dragonflies or to catch a meal for itself.

Evening

In the quiet of the evening,
larger animals begin to search for food.

28 Evening

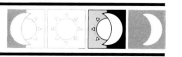

Evening in the forest is a quiet time. Most daytime creatures have hidden themselves away and the nocturnal ones have yet to appear because it is not yet dark.

Some of the larger mammals start to emerge from their daytime resting places. The fox begins its regular route in search of food. Searching for rodents, earthworms and rabbits, it will cover all of its territory throughout the evening and night.

▶ This common shrew has caught an earthworm. After it has eaten, it will sleep for a few minutes and then rush about looking for another meal.

▼ The fox emerges in the evening to find food for its youngsters. It follows regular routes throughout its territory. As it travels, it marks the boundaries of its territory with strong-smelling urine and droppings so that other foxes are aware that the territory is already occupied.

Evening

▼ The garden tiger moth, a fast and beautiful flier, warns off predators with its bright colors. (Inset) The female glowworm is, in fact, a wingless female beetle. The bright light is made by the tip of its abdomen and is used to attract the males so that they can mate. The female can lay up to one hundred eggs.

The last glimmer of daylight fades and the forest now has a different look to it. Dark, menacing shapes begin to emerge as owls and other predatory creatures wake and start their silent search for their first meal of the day.

Small, bright lights of the female glowworm beetles begin to appear in the damp grass around the forest pool. They are always found in the damper areas of the forest, because their larvae only feed on the snails that live there.

Frogs start to crawl out of the water in search of food and small beetles and other invertebrates are active amongst the dead leaves and rotting logs. Millipedes are vegetarian while centipedes feed on slow-moving slugs and snails. Life below the soil surface is just as dangerous as it is above, as each animal has its own predator — most have more than one!

Silently moving through the forest in leaps and bounds, the weasel is probably the most fearsome predator of small mammals. It is a relentless hunter and will even pursue its prey up trees because it is a good climber. It will chase rabbits into the thickest undergrowth and even down into their burrows. The weasel is a tough, strong and much-feared killer and can easily catch rabbits five times heavier than itself.

▲ This millipede is called the pill millipede because it is able to roll up into a ball as a defense mechanism. (Above) This frog has crawled out of the water to look for its first meal of the day. The cooler conditions at night also bring out the slugs and worms on which it feeds.

32 Evening

▼ The weasel is a fearsome predator of the mice and rabbits in the forest.

Night

At the end of the day,
nocturnal creatures, large and small,
start to become active..

Night

By midnight, most of the nocturnal animals are active and small mammals, like the woodmouse, search for food. It will be looking for fruits, seeds and fungi that will then be taken underground to a network of tunnels and sleeping chambers. The minimum amount of time will be spent above ground because weasels and owls will also be out hunting for food.

By the side of the forest pool, the common frog has eaten several slugs and snails since it emerged in early evening. Should it be disturbed, it will leap instantly into the water and hide under the surface until the threat has passed. It can easily jump several times its own length in an effort to escape danger.

▶ The little owl is the smallest owl in the forest. This one is returning to its nest in a hollow tree with a beetle it has caught. Little owls normally only feed on small prey like insects and eathworms, although they can sometimes catch mice and voles.

▶ This frog has been feeding by the side of the forest pool. If it feels threatened in any way, it will escape by leaping into the water and hiding below the surface. As it jumps, it protects its eyes with special transparent eyelids.

◀ At night, the nocturnal woodmouse searches the forest floor for nuts, berries, flowers and fungi to take back to its underground system of tunnels and chambers. There, it will feed on them in relative safety, away from the threat of predators.

38 Night

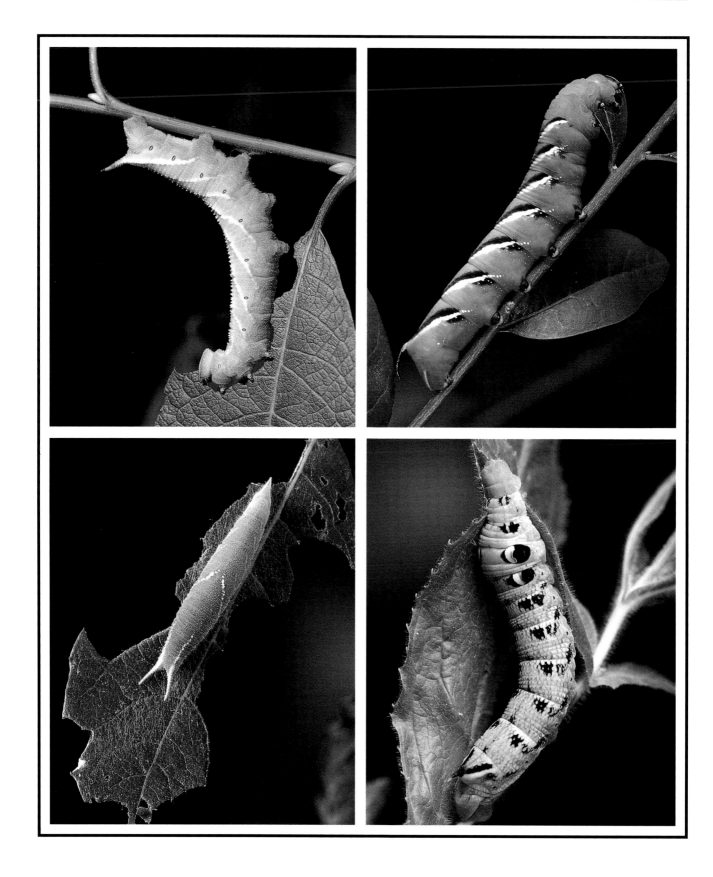

◀ The difference between the purple emperor (bottom left) and the hawkmoth caterpillars can be clearly seen. It is much smaller and does not have a horn on its back. This caterpillar hibernates throughout the winter on the bark of a broad-leaved willow tree which is its food plant. It will pupate and emerge as an adult in one season and will be the most beautiful butterfly in the forest.

There is as much activity in the trees as there is on the ground. Many butterfly and moth caterpillars feed only at night on all kinds of trees and shrubs.

The privet hawkmoth caterpillar, with its fierce appetite, can eat large amounts of food, so the female moth must only lay her eggs at random on each food plant to ensure an adequate food supply.

Six weeks after the eggs have hatched, the fully grown hawkmoth caterpillar climbs down to the ground and digs into the leaves on the floor of the forest. It pupates here and can often stay as a pupa until the following spring. Sometimes, however, it will emerge as the adult moth within three weeks, completing the life cycle within one summer.

▶ The privet hawkmoth is so called because its larvae and caterpillars feed on privet leaves. It is a fast, powerful flier, shaped like a hawk or bird of prey. Its body is steamlined and its wings are long, narrow and pointed. It can fly fast and can also hover when it feeds from sweet, scented flowers.

◀ Hawkmoth caterpillars are also sometimes called hornworms because most species have a horn on their back. When they are fully grown, the hawkmoth caterpillars stop feeding and crawl away until they pupate in the leaves on the forest floor.

Night

At three o'clock in the morning, the badger is making its way back to its sett deep in the forest. For four or five hours it has been walking the regular pathways of its territory looking for food. Sometimes it will travel for miles in search of a good feeding area like a meadow, where it can find a plentiful supply of earthworms.

The tawny owl is the biggest bird in the forest. Its silent flight and sensitive hearing make it a formidable hunter. Every night with its strong, sharp talons it will catch many small mice and voles to feed to its noisy youngsters. When the first dawn light of a new day appears it will retreat to its nest and quietly sleep until night falls again.

▶ The tawny owl nests in a hollow tree in the forest. It has very strong, sharp talons and sensitive hearing and flies silently through the forest in search of small mammals like the woodmouse, shrews and bank voles.

▼ The badger travels long distances in search of food. The badger is a strong digger and is able to dig its sett deep into the forest soil sometimes as much as four meters (13 feet) underground.

42 Night

▼ Just before dawn, if the humidity and temperature are right, the Peziza cup fungus releases a cloud of spores, triggered by a gust of wind or passing mammal.

▶ The coprinus fungus is a small toadstool growing on the forest floor and is often eaten by woodmice and slugs and snails.

Glossary

Aphid
Small insect that sucks and damages trees and other plants. Greenfly are common examples.

Amphibian
Animal that lives on land and in water, for example, frog and newt.

Camouflage
The color and shape of an animal's shell or skin that enable it to blend with its habitat to hide from predators.

Defense mechanism
A physical response by the body whenever danger is suspected.

Elytra
Hard wing cases of a beetle.

Foliage
A mass of leaves and plants.

Glade
An open space in a wood or forest.

Habitat
The normal home or locality of animals and plants.

Hibernate
To go to sleep during the winter months.

Honeydew
Sugary liquid made by aphids.

Invertebrate
Animal not having a backbone, for example, butterfly, slug, earthworm.

Larva
The stage of an insect life cycle in between egg and pupa, for example, caterpillar.

Lichens
A very simple fungus-like plant that grows on trees and rocks.

Nectar
Sweet liquid made by flowers to attract insects.

Nocturnal
Active only at night.

Metabolic rate
The process of burning food within the body in order to grow and live.

Nymph
An immature form of insect which may or may not resemble the adult.

Predator
Animal or insect which feeds on another.

Preen
To clean and arrange feathers or fur on an animal or bird's body.

Proclaim
To announce with authority.

Pupa
Stage of an insect between larva and adult.

Rodent
Small animal with strong incisor (front) and no canine (back) teeth, for example, mouse, squirrel, bank vole.

Rutting season
Period each year, usually late summer, early autumn, when deer gather together to mate.

Scuttle
To move around quickly.

Spawn
A mass of eggs which are laid in water.

Territory
Area of land or forest claimed and used as home by an animal.

Tree canopy
The uppermost layer of branches in a forest.

Index

amphibians 20, 43
aphids 21

badger 40
bank vole 18, 40
beetles 11, 20, 31
blackbird 10

brown speckled wood butterfly 22
butterflies 12, 22, 39

camouflage 23, 43
carpenter ant 21
caterpillars 39
centipede 31
common blue butterfly 22
common frog 20, 28, 31, 36, 37
common shrew 29
coprinus fungus 42

dragonflies 12, 13, 24

earthworms 28, 29, 36, 40
elytra 20, 43

fly 12
fox 18, 19, 28, 29
frog 18, 20, 28, 31, 36, 37
fungi 36, 42

garden tiger moth 23, 30
glowworm 30
gray squirrel 11

hawkmoth caterpillar 38, 39
hornworm 38, 39

insects 11, 12, 36
invertebrates 31, 43

lichen 11
little owl 36, 37

mating 11, 21, 30
mice 32, 36, 40
millipede 31
moths 39

nectar 20, 22
newt 20
nymph 12, 14, 24

owls 30, 36

peacock butterfly 22, 23
peziza cup fungus 42
privet hawkmoth 38, 39
purple emperor 38, 39
pygmy shrew 18

rabbit 28, 31, 32
rodents 28
rosebay willowherb 12
ruddy darter 12, 13, 24
rutting season 11, 43

shrew 18, 29, 40
slugs 31, 36, 42
snails 31, 36, 42
southern hawker dragonfly 14
spiders 14
spores 42
spotted longhorn beetle 20
stag 11

tawny owl 29, 40, 41
territories 10, 28, 40
tortoiseshell butterfly 12

vole 18, 36, 40

weasel 31, 32, 36
wing cases 20, 43
woodmouse 36, 40, 42

PRINTED IN BELGIUM BY
proost
INTERNATIONAL BOOK PRODUCTION